创意无限·设计图典系列

网页设计手册

王凤波　主编

化学工业出版社
·北京·

编写人员名单（排名不分先后）

任 昆　王君飞　马 飞　秦娇娇　邱香宝　白秀英
冷玉岩　陈新刚　魏大海　吕梓源　孙 涛　任俊秀
王凤波　许建伟　张凤霞　许海峰　陈素敏

图书在版编目（CIP）数据

网页设计手册/王凤波主编. —北京：化学工业出版社，
2008.5
（创意无限·设计图典系列）
ISBN 978-7-122-02544-9

Ⅰ.网… Ⅱ.王… Ⅲ.主页制作-图集 Ⅳ.TP393.092-64

中国版本图书馆CIP数据核字（2008）第049947号

责任编辑：王 斌 徐华颖　　　装帧设计：王凤波
责任校对：李 军

出版发行：化学工业出版社
　　　　　（北京市东城区青年湖南街13号　邮政编码100011）
印　　装：北京画中画印刷有限公司
889mm×1194mm　1/32　印张4½　2008年5月北京第1版第1次印刷

购书咨询：010-64518888 (传真：010-64519686)
售后服务：010-64518899
网　　址：http://www.cip.com.cn
凡购买本书，如有缺损质量问题，本社销售中心负责调换。

定　价：29.00元　　　　　　　　　　版权所有　违者必究

编者的话

《创意无限·设计图典》系列包括《VI设计手册》、《CI设计手册》、《宣传册设计手册》、《广告设计手册》、《包装设计手册》、《网页设计手册》、《POP设计手册》、《ICON设计手册》、《装饰纹样设计手册》、《动漫形象设计手册》、《工业造型设计手册》、《时装画设计手册》、《装饰画创意手册》、《插画创意手册》、《版式设计手册》，该系列汇集了大量经过精心挑选的图片资料，列举了国内外优秀的创意作品。这些作品从不同角度、不同表现风格、不同表现方式说明其创意主题，内容涉及文化、商业、工业、传媒等行业的诸多类别。本系列书中收录的各种类型的设计作品视觉冲击力极强，是艺术性和商业性完美结合的优秀作品，为各个行业的设计提供了丰富的范例。

设计艺术发展到今天，已没有什么闻所未闻的技巧可言，在这个领域里，所运用的设计技巧大同小异，同样的技巧、同样的技法很多人都在用，但最终表现的效果却不一样，这就是如何运用的问题。本系列书则是通过大量的范例，以体现技法的运用，显示出作品的风格与内涵；起到激发创作灵感的作用，对专业设计者、在校学生以及业余爱好者有极高的参考价值。

编者

目录

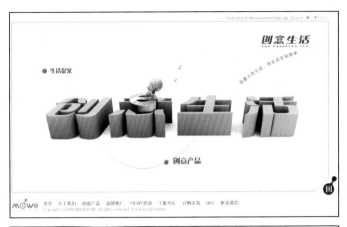

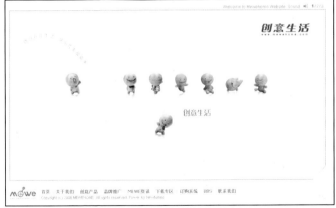

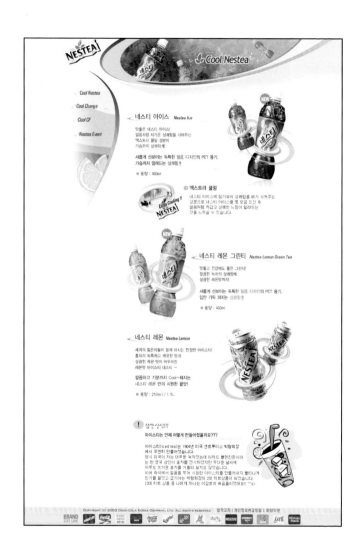

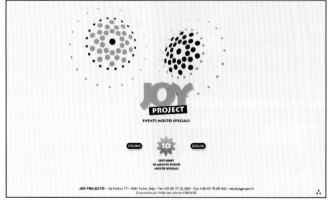

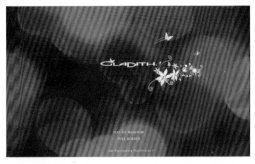

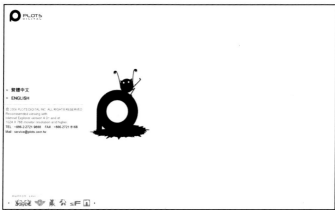

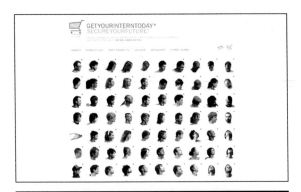

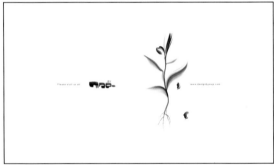

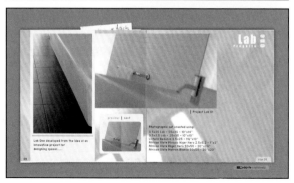

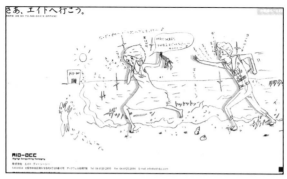

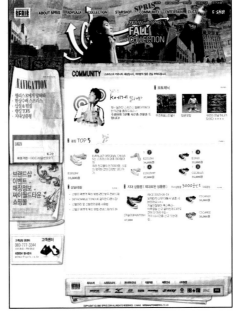

classes. Its famous yellow or tan two-tone grooved sole and signature footprint made it a classic.

Adopted by youth culture as a symbol of self-expression, Dr. Martens later evolved from its humble origins to a brand that would create a different cultural footprint decade after decade.

Now, more than forty years on, a continuous succession of skinheads, mods, punks and various other cultural renegades have adopted the 1460 as a symbol of their independence and individuality whilst giving a nod to the boot's working class roots. Likewise, long list of those who can be counted as icons of style and symbols of rebellion have gone on to help paint Dr Martens rich history. Those such as The Who, The Sex Pistols, The Clash, Madness, The Cure, The Red Hot Chili Peppers, No Doubt - the list goes on.

In the spirit of originality inherent to Dr Martens and in celebration of its past and future, Dr. Martens has asked designers from Britain and around the world to customise its most classic boot. Featured here are images of the original, humble cherry red, eight eyelet 1460 work boot - each with original impressions left by leading designers.

As you would expect of creative talent and original thinkers, the results are nothing short of inspirational, proving that when it comes to individuality and self-expression the Dr. Martens 1460 boot continues to be as iconic as ever.

PURCHASE

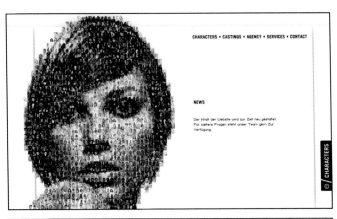

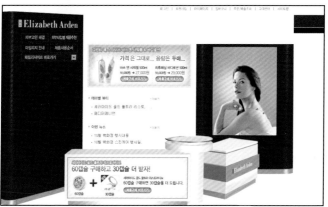

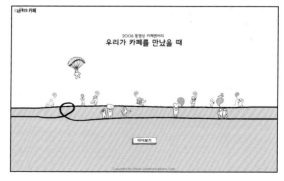

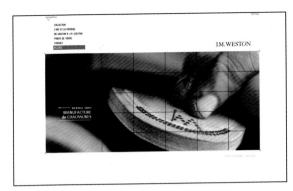

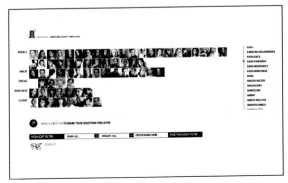

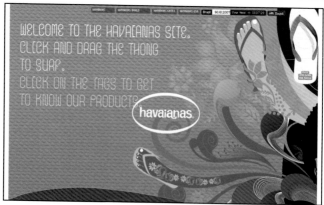

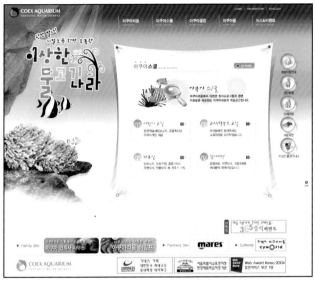

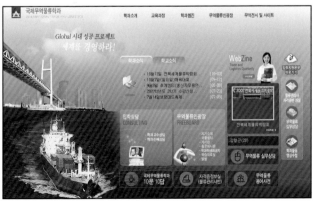

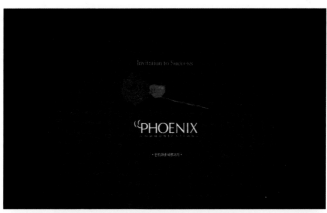

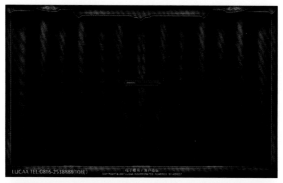

玫瑰之约

NANHU
ROSE
BAY

开启 孕育山水灵气 演绎现代经典
Nanhu Rose Bay

☑ 南湖玫瑰湾（MENU）

Nanhu Rose Bay

孕育山水灵气 演绎现代经典
Gestation of Landscape's Nimbus
Present of Modern Classic

☑ 更多新闻   泰兰庭湖玫瑰湾夏日盛夏江城亮相耀眼光   2007-8-21
         楼盘案名解析系列之泰兰南湖玫瑰湾   2007-8-21
         泰兰武汉项目身地推介—玫瑰湾深探徘徊湃   2007-8-21

027-8738 9999   建筑设计：深圳筑博 ／ 园林设计：加拿大奥雅 ／ 营销顾问：Tao大客顾问司 ／ 整合推广：博思堂广告 ／ 全程网站营销：武汉聚烽网

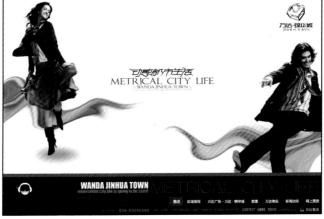

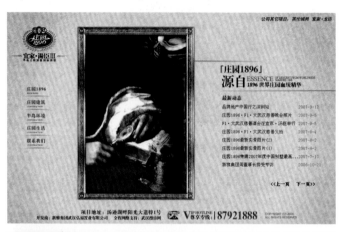

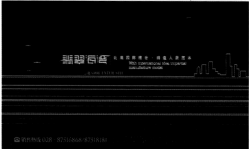

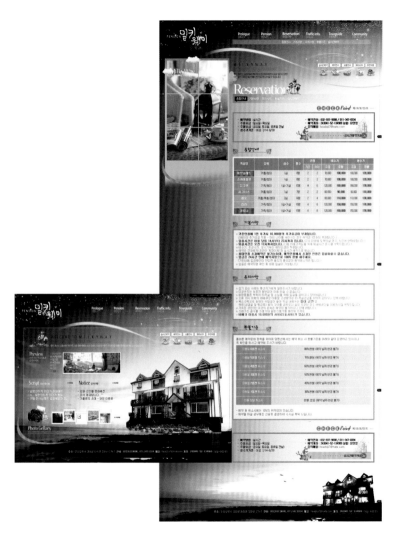

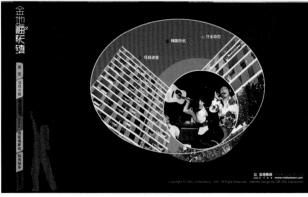

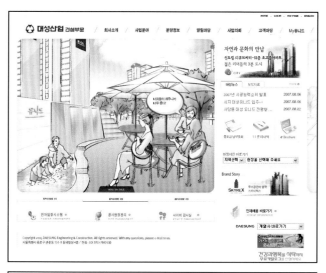

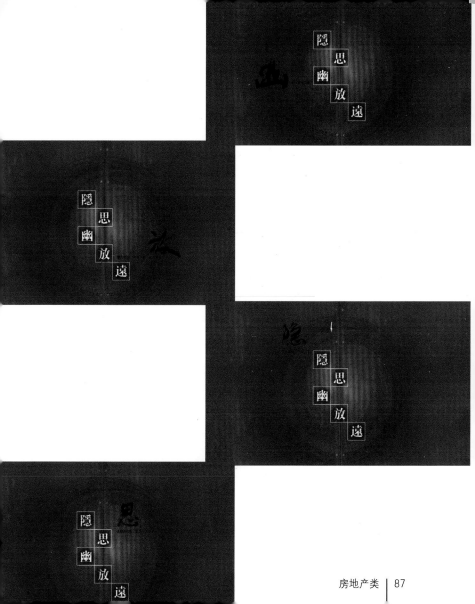

隐
思
幽
放
遠

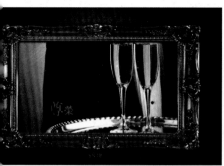

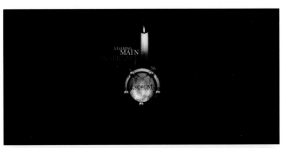

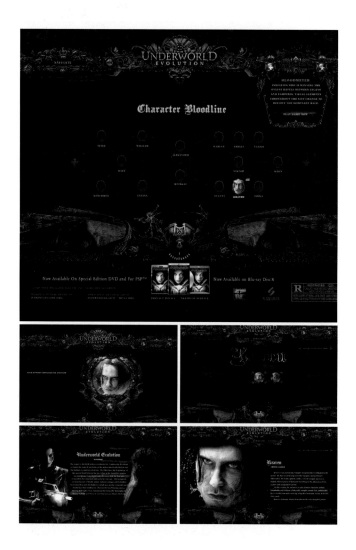

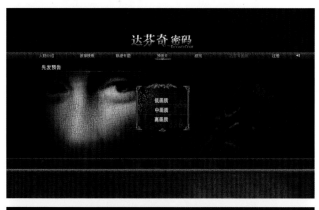

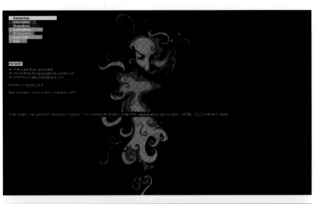

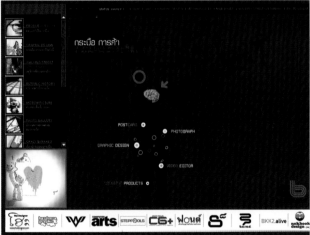

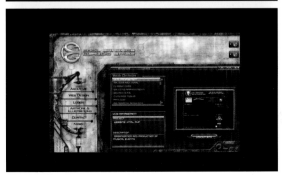

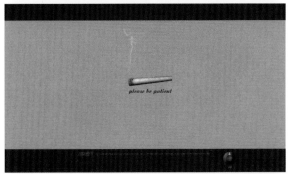

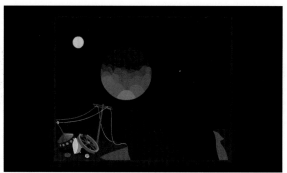

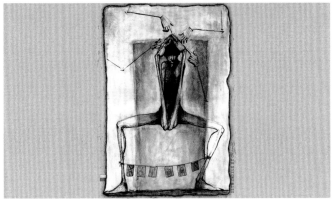

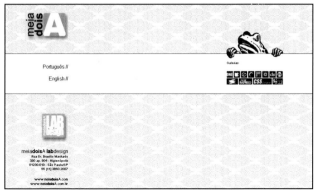

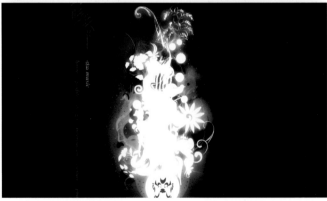

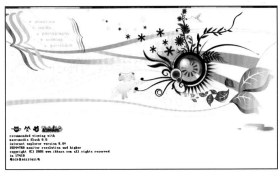

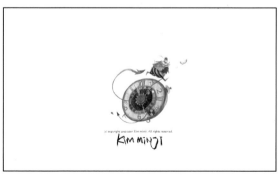

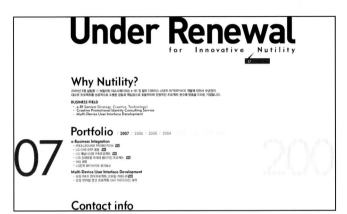

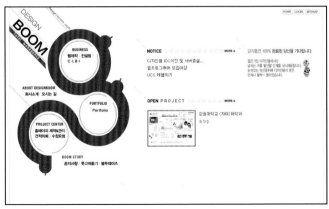

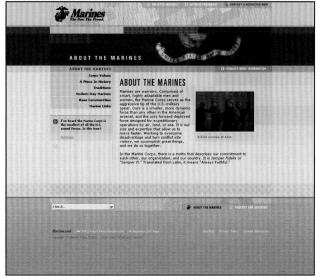

WELCOME TO
THE VODAFONE JOURNEY
JOIN US FOR A LOOK AROUND VODAFONE

 Take the Guided Tour.
FOLLOW OUR GUIDE

 Our Story, Your Story.
MEET OUR CUSTOMERS

 Where You Find Us.
A TRULY GLOBAL COMPANY

 Our Services.
SEE HOW WE ENRICH PEOPLE'S LIVES

 Being Responsible.
FIND OUT HOW WE BUILD TRUST IN OUR BUSINESS

 Looking Ahead.
CHECK OUR PLANS FOR THE FUTURE

Come and have a look around Vodafone. Meet our customers worldwide, explore our present
and our future, and experience the impact we have on people's lives.

 You enjoyed this journey — so share it!
SEND THE VODAFONE JOURNEY TO A FRIEND

LOADING CONTENT FOR
THE VODAFONE JOURNEY
PLEASE WAIT   FOR BROADBAND CONNECTION (35 SECONDS)

55%

Come and have a look around Vodafone. Meet our customers worldwide, explore our present
and our future, and experience the impact we have on people's lives.

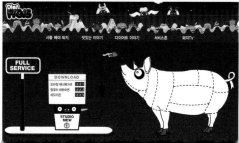

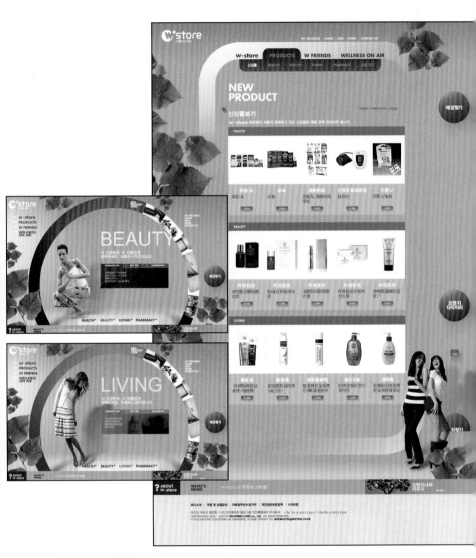

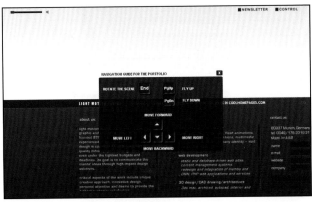

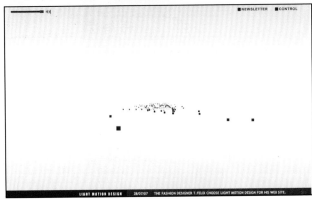